NIST Technical Note 1811

Applying the Hedonic Method

Stanley W. Gilbert
Office of Applied Economics
Engineering Laboratory

September 2013

U.S. Department of Commerce
Penny Pritzker, Secretary

National Institute of Standards and Technology
Patrick D. Gallagher, Under Secretary of Commerce for Standards and Technology and Director

Abstract

Often, advances in measurement science will enable the development of new product innovations. When trying to estimate the value of the benefits of an advance in measurement science, it can be difficult to estimate the value of those new product innovations, especially if they are contained within composite goods.

The hedonic method provides a mechanism for the impact on the price of a composite good of a new product innovation. Under the right circumstances, and with the right data, it can even help estimate the value to the consumer of those new product innovations. This helps in estimating the benefits of advances in measurement science, and in some cases can help show where additional resources could be profitably spent in seeking further advances.

This Technical Note describes what the hedonic method does, how one is done, what its limitations are. It starts with a discussion of the theory behind hedonic studies, including a discussion of the basic assumptions behind it. It then goes on to describe how such studies are done. It specifically discusses basic studies that simply estimate the price function for a composite good, including some of the potential problems that could arise in such a study. It then goes on to discuss more complete studies that estimate supply and or demand functions for the market for a composite good. It concludes with an annotated bibliography.

Keywords: Hedonic Method; Hedonic Bibliography; Composite Goods; Theory and Practice

Acknowledgements

The author wishes to thank all those who contributed so many excellent ideas and suggestions to this report. They include Dr. Robert Chapman, Chief of the Applied Economic Office (AEO) at the National Institute of Standards and Technology who provided support and many useful comments. Dr. David Butry, an AEO economist, provided many useful comments. Ms. Shannon Takach of the AEO assisted greatly in preparing the manuscript for review and publication. As always, any remaining flaws and errors are the responsibility of the author.

Disclaimer regarding Non-Metric Units

The policy of the National Institute of Standards and Technology is to use metric units in all of its published materials. Because this report is intended for the U.S. construction industry that uses U.S. customary units, it is more practical and less confusing to use U.S. customary units rather than metric units. Measurement values in this report are therefore stated in U.S. customary units first, followed by the corresponding values in metric units within parentheses.

Contents

Abstract iii

Acknowledgements v

Introduction 1

Theory 3

    Basic Assumptions 4

        Completeness 4

        Availability 4

        Market Power 4

    Preference / Cost Estimation 5

Estimating the Price Function 6

    Assumptions 6

        Functional Form 6

        Characteristics included in the Analysis 7

        Spatial Correlation 8

        Techniques to Address These Issues 9

Estimating Preferences 9

Conclusion 10

Annotated Bibliography 13

    Theory of Hedonic Analysis 13

    Preferences and Costs 15

    Price Function 18

Introduction

A hedonic study is an effort to understand what characteristics contribute to the value of a good. Most goods have a variety of characteristics that contribute to or detract from its value. For example, the characteristics of a ballpoint pen that contribute to (or detract from) its value would include the color of ink, durability of ink, how fast the ink dries, how fine a point it has, how big it is, how readily it leaks (for example, does it tend to leak on airplanes), how easy the ink is to refill, and how stylish it is. Goods like this are called Composite Goods.

A Composite Good is a good that consists of a bundle of characteristics that are relevant to people's decision to buy them. Examples of composite goods include houses, cars, phones, and food. A car, for example, has a number of characteristics that factor into how much people are willing to pay for it. These include color, body style, seating capacity, fuel efficiency, and horsepower. The purpose of a hedonic study is to determine how much each characteristic contributes to the value of the car at the margin. For example, Goodman (1983) found that for cars manufactured in 1975 and sold on the used-car market in 1977, a one MPG increase in fuel efficiency increased the resale price by $93.

Hedonic studies are primarily useful for estimating the value of non-market goods where those non-market goods form part of some larger composite good. Examples of studies that have done this include studies that estimated the value of fire sprinklers in single-family residential houses, energy efficiency in air conditioners, the value of trees in a residential neighborhood, and the value of brand names for laser printers. This sort of a study could be useful in developing cost-benefit analyses of proposed seismic or fire design codes (where, for example, code changes could eliminate (or enable) building features or building types that people value for their own sake); it could aid in developing cost-benefit analyses of low- (or zero-) energy-use structural designs; or it could enable a value to be placed on improved building comfort when (for example) commissioning buildings.

A Hedonic study can have at least one of two basic objectives. First, and most commonly, hedonic studies seek to define the relationship between the characteristics that make up the Composite Good and its price. Second, hedonic studies sometimes also attempt to describe the individual preferences that drive people's decisions about what they buy. The latter is much more difficult to do correctly, and so tends to be done only rarely. The classic example of a preference study is Witte, et al. (1979).

The basic approach to a pricing study is illustrated by Rao and Lynch (1993) in their hedonic study of computer workstations[1]. We assume that the products in the market being studied are completely described by some vector Z of characteristics. Thus any particular product, i, will be defined by its bundle of characteristics, z_i, and will sell for price p_i. Then, the basic assumption of any hedonic pricing study is that there is a function, p, that determines the price of all goods in the market, where $p_i = p(z_i)$.

There are two basic decisions that need to be made before the study really commences. First, is the decision of what characteristics to include in the description of goods in the market. Typically no claim is made that the characteristics selected provide a complete description of the Composite Good. The second decision that must be made is what functional form to use in modeling the pricing relationship.

---

[1] Workstations were an intermediate class of computer between the standard desktop microcomputer and the mainframe.

In the Rao and Lynch (RL) study, they finally settled on six variables describing the system, and three dummy variables representing the manufacturer (see table). The Manufacturer dummy was included because it was believed that the reputation for quality and the level and quality of service provided contributed to the value of the system for buyers.

RL made an attempt to determine the functional form that best fit the data. Based on those efforts, they settled on a linear specification. That is, they assumed that price was related to the characteristics by the following function:

$$p_i = \beta_0 + \beta_1 DASD_i + \beta_2 RAMLOW_i + \beta_3 MONO_i + \beta_4 SCSI_i + \beta_5 MIPS_i$$

$$+\beta_6 GRAPH_i + \beta_7 SUN_i + \beta_8 DEC_i + \beta_9 HP_i + \varepsilon_i$$

Here the variable ε is a random variable accounting for the fact that they didn't actually have a complete specification of the Composite Good.

At that point, RL simply regressed price against the workstation characteristics using publically available pricing data. Based on their regression, they found that the variables with the greatest effect on workstation price were DASD, SCSI (which negatively affected price), and RAMLOW.

Many studies simply assume that the results of the price study are a direct reflection of underlying consumer preferences. It must be emphasized that this is not necessarily the case. First, price is a product of both supply and demand. Factors that affect the supply side of the market will be reflected in the price even if they have no impact whatsoever on consumer preferences. For example, the DEC manufacturer variable was strongly negative. One possible explanation is that customers considered DEC machines to be of lower quality than those of other manufacturers, or that

Variable	Description
DASD	Maximum size hard drive (or other direct-access secondary storage) supported.
RAMLOW	Minimum amount of RAM required.
MONO	Dummy Variable: Does the machine support a monochrome monitor?
SCSI	Dummy Variable: Does the machine use a SCSI mass storage disk controller?
MIPS	A measure of system performance.
GRAPH	Number of graphics standard supported.
SUN, DEC, HP	Manufacturer dummies.

consumers found the service package provided by DEC to be less valuable than that of other manufacturers. However RL attributed the price difference to supply differences: specifically they observed that DEC had been the first manufacturer to significantly lower its price in search of market share, thus setting off a price war in the workstation market.

Second, it is likely that consumer preferences are heterogeneous. In such a case, a simple description of preferences becomes difficult at best. For example, in the RL study, the SCSI variable is strongly negative. RL suggest that the SCSI variable marks the difference between a standard workstation which came equipped with a SCSI Bus and special purpose workstations which often did not, and tended to be much more highly priced. In this case the SCSI variable reflected a divide not only between different types of workstations, but also between different types of customers who valued different characteristics in the products they were buying.

This report describes what the hedonic method does, how a hedonistic study is done, what its limitations are. It starts with a discussion of the theory behind hedonic studies, including a discussion of the basic assumptions behind it. It then goes on to describe how such studies are done. First it discusses how a basic study that simply estimates the price function for a composite good is conducted, including some of the potential problems that could arise in such a study. It then goes on to describe how a more complete study that estimates supply and or demand functions for the market for a composite good would be conducted. It concludes with an annotated bibliography.

Theory

One of the earliest treatments of the idea of a Composite Good was Lancaster (1966). He developed a consumer theory where all goods are composite, and people buy those goods for the characteristics that make them up rather than for the goods themselves. However, the seminal treatment of the idea, and the one that placed the Hedonic study on firm footing was Rosen (1974). His model assumes that both consumers and producers are price takers (i.e., they must both take the market price). He then derives supply and demand from the consumers' preferences and from the producers' profit-maximization decision.

Assume that there exists a price function $p(z)$, where z is a vector of characteristics completely describing the Composite Good being studied. That is, the characteristics in z completely determine the price paid every time a consumer buys one of the Composite Goods being studied. Also assume that consumers have utility functions $U(c, z; x)$, where c represents consumption of all other goods, and x parameterizes different consumers. Note that this model explicitly allows different consumers to have different preferences. Each consumer chooses the version of the Composite Good to purchase in order to maximize his or her utility, taking into account price. That is, the consumer solves the problem:

$$\max_{z} U(y - p(z), z; x)$$

Here y represents the consumer's income.

Similarly for producers, we have a cost function, $C(M, z; \beta)$, where M is the amount produced, z is the vector of characteristics of the good produced, and β is a vector of characteristics that define the different producers. Again, the model explicitly allows different producers to have different cost structures. Each producer selects the specific Composite Good (z) to produce and how much to produce. That is, the producer solves the problem:

$$\max_{M,z} Mp(z) - C(M, z; \beta)$$

From these we can derive supply and demand for each z. Specifically $Q^D(p, z)$ would be the number of consumers who choose to consume bundle z given the price function p, and $Q^S(p, z)$ would be the total amount of good z produced given the price function p. Supply and demand must be equal for all values of z. The price, then, is the function that results in $Q^D(p, z) = Q^S(p, z)$ for all z. Note that this will typically require the solution of a highly non-linear second-order differential equation.

3

In practice, we aren't interested in solving that differential equation. In fact, the data we typically have generally allows us to estimate $p(z)$ directly. What we are interested in doing, at least some of the time, is solving the inverse problem: that is, trying to estimate the utility function(s) and / or the cost function(s) from the pricing data.

The fundamental assumption that any hedonic study makes is that a stable price function $p(z)$ exists. This in turn implies an additional number of assumptions that typically include:

- *Completeness*: All possible products within the product space are available for sale.

- *Availability*: At any given time within a single market, all products are available to all consumers—that is, we are looking at a single unified market.

- *Market Power*: No consumer or producer has market power—that is, all participants in the market are price-takers.

As a practical matter, of course, these assumptions are often violated in practice. How serious an impact these violations have on the study depends on the situation. With that in mind, each of the assumptions is discussed in turn.

Basic Assumptions

Completeness

Completeness mainly ensures that each consumer gets his or her optimal bundle. If it does not hold, then consumers will tend to "bunch" up on the boundary of their choice set. That will be reflected in the estimated price function. If not properly accounted for it could easily lead to misinterpretations of the results.

In practice, this condition is rarely fully met. How serious the problem is depends on how incomplete the market is, and what form the market incompleteness takes. In some cases, the incompleteness can be safely ignored (and it often is ignored in housing studies). However, if significant portions of the product space are simply not available for sale, then boundary effects may need to be taken into account. Certainly extrapolation beyond the bounds of the market is likely to be seriously problematic.

Availability

Availability matters because if some non-negligible segment of the population has no access to some of the products on the market, then prices will be too low for the inaccessible market segment, and likely too high for the remainder of the market. Remember that prices are a product of supply and demand. If demand is artificially restricted for some segment of the market, then prices will partially reflect that restriction.

Market Power

Market power matters because interpretation of the price function can become seriously problematic when some party (or parties) have pricing power. For example, in a monopoly market, price is determined by the monopolist. That enables complications like price discrimination (where different customers are charged different prices for the same good) and strategic pricing decisions on the part of the monopolist (for example, the monopolist might charge higher prices for one good in an effort to channel customers to a related but more profitable good).

This assumption is somewhat less likely to be violated in practice, but is still violated from time to time. It is reasonable to assume that housing markets fit this assumption. An example of a study where the assumption was violated is the RL study above. In it three companies controlled 70 % of the market, suggesting that they had some pricing power. The study also makes it clear that price discrimination was routinely practiced in the market. The prices used in the study were the list prices, not the actual prices paid by customers. Since price discrimination was routinely practiced, the list prices likely had a significant strategic component to them.

If some market participant has market power, then care needs to be taken in interpreting the data. Furthermore, since pricing is a strategic decision, then predicting what will happen as a result of market changes becomes much more difficult.

Preference / Cost Estimation

As previously discussed, the results of a pricing study need to be interpreted carefully. An example provides an additional illustration of this. Goodman (1983) attempted to estimate the value of fuel efficiency for cars on the U.S. used car market. He specifically looked at a select subset of models, sold on the used car market in 1977 and 1979, and that were two years old at the time. In 1977 he found that a one MPG increase in fuel efficiency increased the resale price of 1975-model cars by approximately $93 (or $219 for a one km / l increase). However, in 1979, he found that an increase in fuel efficiency *reduced* the resale price of a 1977-model car. In spite of valiant effort, he was unable to overturn that conclusion. That result is counter-intuitive, since we would expect fuel efficiency to have a positive value for consumers (as further indicated by the 1977 results).

However, supply-related factors may serve to explain his results. The Corporate Average Fuel Economy (CAFE) standards came into force in 1978 and were significantly tightened for the 1979 model year. Those standards require auto makers to maintain a minimum average fuel economy for the automobiles sold in a specific year. Assuming (as seems likely) that the standards mandated a mix of cars different from the one that consumers would have purchased on their own in 1979, then low-fuel-efficiency new cars would have carried an additional cost for complying with the CAFE mandate. Assuming that new and used cars are to some extent substitutes, and since the increase in price did not affect demand, the "overpricing" of low-fuel-efficiency new cars would to some extent have been compensated for by a similar "overpricing" of low-fuel-efficiency used cars.

A number of techniques have been discussed and used to estimate utility parameters or cost parameters from the pricing data. Rosen (1974) suggested one approach. To implement that technique required data from multiple separate markets. Essentially, he required multiple instances where supply and demand were different. These could be geographically distinct markets, or it could be the same geographic market observed at different times when supply and / or demand would have shifted. Then, given the pricing function (which is estimated first), and the marginal price (determined from the estimated pricing function), he suggested simultaneously estimating the system of equations representing supply and demand.

As Brown and Rosen (1982) discuss, it is essential that multiple markets be used in this technique, or that prices be non-linear. They observe that if marginal prices are linear, then the method described above produces no information about the underlying supply and demand functions. On the basis of their results, they argued that multiple markets were imperative if an attempt was to be made to recover structural information about supply and demand.

Ekeland et al. (2004) soften this conclusion by noting that a non-linear pricing structure will allow structural information to be recovered. They also note that marginal price is "almost always" non-linear (where "almost always" had a technical definition that is beyond the scope of this paper). Nevertheless, the key observation of Brown and Rosen holds: care must be taken when trying to estimate structural parameters, or the results will be meaningless.

Estimating the Price Function

Assumptions

As a practical matter, when trying to estimate the price function $p(z)$, all the factors that affect the price are not known. So, in practice, the related function, $p(z, \varepsilon)$, is what is estimated, where z are those characteristics of the good that are known to the researcher, and ε are those characteristics that are not known. In particular, it is assumed that the unknown characteristics can be described by a single random variable that is distributed according to some probability distribution $F(\varepsilon)^2$. Estimation of the price function then is the process of estimating $p(z, \varepsilon)$ accounting, as necessary, for the probability distribution $F(\varepsilon)$. In some cases the parameters of the probability distribution, F, are also estimated.

Functional Form

As with any statistical analysis, there are a number of decisions that must be made. In estimating the price function, the same considerations apply that apply to any regression model. First and foremost is the functional form to be estimated. As a general rule most studies assume a linear relationship between the price and some function of the characteristics, or some form that can be transformed to linear. Log-log models and log-linear models are common variants.

The Box-Cox transformation (which includes linear and log models as special cases) is used somewhat less frequently. The Box-Cox transformation is shown below.

$$x_{(\lambda)} = \begin{cases} \dfrac{x^\lambda - 1}{\lambda} & \lambda > 0 \\ \ln x & \lambda = 0 \end{cases}$$

Some researchers have used the Box-Cox transformation and estimated the optimal exponent as part of the regression (including Goodman, 1983 and especially Halstead et al. 1997).

Non-Linear models are also occasionally estimated. For a particularly complex example see Epple et al. (2006). Many non-linear models can be readily estimated using modern software and methods;

---

[2] More complex assumptions can be made, but that is rarely done.

however, non-linear models are rare because they are more complex to set up, convergence issues are more likely to crop up, and generally they require an *a priori* justification for their form.

Functional form is important because an incorrect form can produce incorrect results. How large the error is in the results will depend on the problem but could be quite large.

Characteristics included in the Analysis

Which characteristics to include in the description of the composite good is also important. Two problems are considered: inclusion of irrelevant characteristics and failure to include relevant characteristics.

Inclusion of irrelevant characteristics will not generally bias the results of the estimated price function; however, it can reduce the significance of the results obtained or produce spurious results through simple random variation. An example will illustrate the latter problem. Kochi et al. (2012)[3], estimate the mortality impact of urban air pollution from the 2003 Southern California wildfires. They conducted a total of 36 regressions for twelve different portions of the study area and three different time periods during and following the fire. They found significant increases in mortality in three of the regions studied. One region was found to have increased mortality during the fires (at the 5 % level). Two other regions were found to have increased mortality during the two weeks following the fire (one at the 10 % level and one at the 1 % level). The problem is that these results are difficult to distinguish from random variation. Looking exclusively at the results for mortality during the fire, the likelihood of producing at least one result with a 5 % level of significance out of twelve regressions is about 45 % (their conclusions for the follow-up period were somewhat stronger).

If some of the irrelevant variables correlate with the relevant characteristics, then significance levels for the results will dramatically decline. In short, if an irrelevant characteristic is included, results will still be unbiased, but they may become confused as significance levels decline or spurious correlations start to appear.

Exclusion of relevant characteristics can be more serious. When an omitted characteristic significantly affects price and is correlated with an included characteristic then the results for the included characteristic will misrepresent its impact on price. For example, Bayer et al. (2008) estimated the value that people place on living in a good quality school district. Since good quality school districts tend to be associated with good "quality" neighborhoods, there is a correlation between district quality and neighborhood quality. Not accounting for neighborhood quality will result in an overestimate of the value that people place on being in a good quality school district. Bayer at al. use boundary discontinuity techniques by assuming that neighborhoods that are very close together but on opposite sides of a school district boundary will be similar in quality. That allows them to separate the effects of neighborhood quality from school district quality.

Related are situations where a characteristic is simultaneously determined with price. The classic example of this problem is the simultaneous determination of quantity and price of some market good. In the classic model, both the quantity and price sold of a market good are determined by the point where supply and demand are equal. That model can be described using the following set of simultaneous equations:

[3] Kochi, I., P. A. Champ, J. B. Loomis, and G. H. Donovan. 2012. "Valuing Mortality Impacts of Smoke Exposure from Major Southern California Wildfires." *Journal of Forest Economics* 18 (1): 61–75.

$$Q = Q_D(P, X)$$
$$Q = Q_S(P, Y)$$

By the Implicit Function Theorem, those equations can be recast as:

$$P = P(X, Y)$$
$$Q = Q(X, Y)$$

That allows us to estimate either price or quantity separately using the techniques described above. This is just restatement of the results of Rosen (1974) in a different form. However, if, in trying to estimate price, we include quantity as one of the regressors (and, as usual, use only a subset of the characteristics); that is, if we try to estimate:

$$P = \hat{P}(Z, Q, \xi)$$

Where $Z \subset X \cup Y$ (and as usual we assume that the excluded variables can be summarized by ξ with a probability distribution $F(\xi)$) then we will run into problems. Specifically, the coefficients on Z will be biased, and will not represent the true effect of those characteristics on price. In addition, interpreting the coefficient on the endogenous regressor is difficult at best.

Fuerst and McAllister (2011) investigated the effect of eco-labeling on rental rates, sales prices, and occupancy for office buildings. Specifically, they estimated the rental and sales price premiums from an Energy Star rating and a LEED rating on office buildings. They found that either rating produces a significant increase in rental rate and sales price. Dual certification provides an additional premium. They found that an Energy Star rating significantly increased occupancy, while an increase in occupancy for the LEED rating could not be confirmed. However, in estimating rental rates, they included occupancy rate as a regressor, when the two are simultaneously determined through equilibration of supply and demand. That casts doubt on their remaining results.

Spatial Correlation

One common example of an important omitted regressor involves spatial correlation in housing markets. In spatial correlation, the price of two nearby houses are correlated even after taking into account the known characteristics of the houses. Remember that the objective is typically to estimate price using some functional form similar to:

$$y_i = x_i'\beta + \varepsilon_i$$

Spatial correlation can occur in three different (non-exclusive) forms. First, the omitted characteristics could be correlated. That is, for two different (but "nearby") houses, ε_i could be correlated with ε_j. In the housing example, that would be the equivalent of having housing characteristics not accounted for in the regression more likely to be similar if the houses are nearby than if they are far apart. A second form would be if the price of a house depends on the characteristics of houses nearby. An example would be if a new house increases the price for all houses in the neighborhood regardless of their age. A third form would be if the price of a house depends on the price of the nearby houses.

The first form of spatial correlation does not introduce any bias into estimates of the regression coefficients, although it means that it takes larger samples to get reasonable results. Either of the remaining forms of spatial correlation, if present, will bias the results, and estimated coefficients on the regressors will generally misstate the relationship between them and price.

Techniques to Address These Issues

In all of these cases, there are techniques available for dealing with these problems. In the case of an omitted correlated variable, instrumental variables techniques are available for addressing the problem. As an example, Palmquist (1984) used instrumental variables in a multi-market model to estimate the underlying demand for housing. When dealing with endogenous regressors, simultaneous equation methods allow for the estimation of the endogenous regressors at the same time as the quantity of interest. Alternatively, instrumental variables methods will also work with endogenous regressors.

A set of techniques have also been developed to handle spatial correlation. As an example, Hui et al. 2007 and Hui et al. 2012 investigated the impact of a number of environmental and neighborhood amenities on the price of housing in Hong Kong. The amenities evaluated included view, the presence of nearby greenbelt areas, air quality and noise level. Their studies explicitly took spatial correlation into account. Hui et al. (2007) separately estimated models where the omitted characteristics were correlated (i.e., a Spatial Error Model), and where price depends on the price of nearby houses (i.e., a Spatial Autocorrelation Model). Hui et al. (2012) estimated a model where two forms of spatial correlation were present simultaneously. The parameters for the spatial regression were significant for both studies.

Estimating Preferences

When preferences or cost functions are the object of the study, a number of techniques have been developed. The techniques depend on one of two approaches. The first requires estimation of price across multiple markets. The second requires the price function to be non-linear. Both approaches are discussed below. Brown and Rosen (1982) made the point that without one or the other, no information can be obtained about underlying preferences or production cost functions.

One of the earliest studies that estimated underlying preferences was Witte et al. (1982) who estimated preferences for rental housing and the cost function for provision of rental housing in non-metropolitan cities in North Carolina. They used price information from four geographically distinct markets in the state to estimate supply and demand parameters. They assumed that there was a single parameterized utility function and cost function (with the parameters describing the difference in costs between different producers, as well as the difference in preferences between different consumers). First, they estimated the price function for each of the markets they analyzed. From that, they obtained marginal price information. Then they used the price information from multiple markets to estimate underlying preferences using standard multiple-equation techniques. Their supply and demand functions had the expected slope: that is, demand declined as its price increased and supply increased as its price increased. In demand, most parts of the composite good that formed the house were complements: that is, an increase in one good increased demand for the others that make up the house.

Palmquist (1984) estimated demand for housing without using the simultaneous equation techniques used in Witte et al. (1982). He used multiple markets as well. Specifically he used six mid-level metropolitan areas from across the country. He also estimated a price function separately for each city, and used the marginal prices computed from that to estimate demand. In order to consistently estimate demand, he used a set of instrumental variables to correct for the inherent endogeneity of price. He found results that made sense economically (higher prices resulted in lower quantity demanded for housing characteristics).

A slightly different approach was used by Kristofersson and Rickertsen (2004 and 2007). They estimated the demand for fish on Iceland's fish auction markets. In their case, the customers whose preferences make up demand are fish processing firms. Their study used price information from multiple markets as well, except in their case the multiple markets are auctions that occur on different days. They assumed that supply in each daily market was inelastic, since the fish on each days market had already been caught. That allowed them to avoid a simultaneous equations model. They also developed a random-coefficients model for demand. In a random-coefficients approach, instead of estimating regression parameters, the researcher estimates a probability distribution over the parameters. They found that price per unit weight for Cod increased as the fish got larger. They were also able to estimate how much the price per unit weight declined per day of storage or if the fish was not gutted.

As an alternative to the multiple markets approach, preferences can be estimated if the price function is nonlinear. Ekeland et al. (2004) argue that the price function will almost always be non-linear, where "almost always" had a technical definition that is beyond the scope of this paper.

Epple et al. (2006) implicitly used the non-linearity of price to estimate preferences for higher education, as well as the cost-function for providing higher education. Their complex technique involved a two-loop iterative procedure. The outer loop was a maximum likelihood estimation of parameters for preferences and costs. The inner loop used the candidate parameters for preferences and costs to solve for the equilibrium price for all the different levels of educational quality in the model. Their model gave rise "to a strict hierarchy of colleges that differ by the educational quality provided to the students." Based on their model they were able to "simultaneously [predict] student selection into institutions of higher education, financial aid, educational expenditures, and educational outcomes." However, the model was very computationally intensive. They had to use bootstrapping to estimate errors and each bootstrap iteration took 24 hours on a PC to compute.

While such a complex model is a possibility in some cases, there are other techniques available that exploit non-linearity without resorting to such computationally intensive methods. Ekeland et al. (2004) suggested the use of instrumental variables to allow for consistent estimation of preferences. They showed that the variables that parameterize preferences or the cost function will serve adequately as instrumental variables for the estimation.

Conclusion

Often, advances in measurement science will enable the development of new product innovations. When trying to estimate the value of the benefits of an advance in measurement science, it can be difficult to estimate the value of those new product innovations, especially if they are contained within composite goods.

The hedonic method provides a mechanism to estimate the impact on the price of a composite good of a new product innovation. Under the right circumstances, and with the right data, it can even help estimate the value to the consumer of those new product innovations. This helps in estimating the benefits of advances in measurement science.

In some cases a hedonic study can help focus efforts where additional resources would most profitably be spent. Consider the hypothetical example where a composite good has one characteristic with a very high marginal price. Any innovation that makes that characteristic cheaper to produce or integrate into the good will make people better off, either by enabling more of the characteristic to be integrated into the good or by allowing the same amount to be supplied more cheaply.

For example, Deligiorgi et al. (2007) estimated the price effects of different characteristics of broadband service in Europe. They found that the characteristic that most strongly affected price was download speed. That implies that research and development resources would most productively be funneled primarily toward increasing download speeds, where the benefit is greatest.

Annotated Bibliography

This annotated bibliography is divided into three sections. The first is a list of papers that address the theory of hedonic analysis. The second section lists papers that attempt to identify supply or demand (preferences or cost functions) for the market analyzed. The third section lists papers that focus on identifying the price function for the market analyzed.

Theory of Hedonic Analysis

Brown, James N., and Harvey S. Rosen. 1982. "On the Estimation of Structural Hedonic Price Models." *Econometrica* 50 (3) (May): 765–768.

> This paper is interested in the limits of studies that attempt to identify demand and / or supply or other structural characteristics. They observe with an example that if the marginal price function is a linear combination of the terms of the supply and demand functions, then the attempt to estimate structural characteristics will provide no information about those characteristics. Therefore, if information about structural characteristics is sought, multiple markets are essential, and / or, the marginal prices for characteristics must be non-linear.

Ekeland, I., J. J. Heckman, and L. P. Nesheim. 2004. "Identification and Estimation of Hedonic Models." *Journal of Political Economy* 112 (1): 60–109.

> This study supports the conclusions of Brown and Rosen (1982) that in order to obtain information about structural characteristics, multiple markets are essential, and / or, the marginal prices for characteristics must be non-linear. They differ from them in that they argue that marginal prices are non-linear in "almost all" markets (where "almost all" has a technical definition beyond the scope of this paper). Assuming that marginal prices are non-linear, they argue that it is possible to get to preferences even in single-market by exploiting the non-linearity. They propose a couple of techniques to do so, including proposing a set of instrumental variables that will work.

Lancaster, K. J. 1966. "A New Approach to Consumer Theory." *The Journal of Political Economy* 74 (2): 132–157.

> This is one of the earliest treatments of a Composite Good. He developed a consumer theory where all goods are composite, and people buy those goods for the characteristics that make them up rather than for the goods themselves.

Rosen, S. 1974. "Hedonic Prices and Implicit Markets: Product Differentiation in Pure Competition." *The Journal of Political Economy*: 34–55.

This is the seminal treatment of Hedonic theory. He attempts to get information about preferences from market data. His main observation is that the normal form of Hedonic analysis does not do that. He goes on to suggest an approach that will get preferences provided sufficient data of the right type is available.

What is typically estimated in Hedonic analysis is the price functional. He notes that the price functional cannot be (correctly) interpreted as a reflection of underlying preferences. For example, suppose there is only one type of seller, but many types of consumers. In that case, the differences in price will reflect different types of consumers rather than giving the price v. attribute function for any consumer.

Preferences and Costs

Bayer, Patrick, Fernando Ferreira, and Robert McMillan. 2007. "A Unified Framework for Measuring Preferences for Schools and Neighborhoods." *Journal of Political Economy* 115 (4) (August): 588–638.

> This paper estimates household preferences for school district quality and neighborhood quality, where the two are highly correlated. In order to disentangle the effects of school district from those for neighborhood quality, they use boundary discontinuity techniques. Specifically, they assume that neighborhoods on the opposite sides of a school district boundary will be very similar. Otherwise, they use techniques similar to those described in Rosen (1974) to estimate structural parameters.

> They also attempt to determine preferences for some of the characteristics that make up neighborhood and school-district "quality." For example, they find that "households are willing to pay less than 1 percent more in house prices when the average performance of the local school increases by 5 percent." They also find that "there is considerable heterogeneity in preferences for schools and neighbors, with households preferring to self-segregate on the basis of both race and education."

Chintagunta, P., J. P. Dube, and K. Y. Goh. 2005. "Beyond the Endogeneity Bias: The Effect of Unmeasured Brand Characteristics on Household-level Brand Choice Models." *Management Science* 51 (5): 832–849.

> This study develops a theoretical model of brand choice, where they estimate the distribution over preferences. In particular they account for endogeneity in prices resulting from un-accounted-for covariates by using an instrumental variables approach.

> They use the model to estimate preferences for margarine in Denver.

Elrod, Terry, and Michael P. Keane. 1995. "A Factor-Analytic Probit Model for Representing the Market Structure in Panel Data." *Journal of Marketing Research* 32 (1) (February): 1–16.

> This paper develops procedures for estimating preference parameters from market data. Specifically they assume that preferences are randomly distributed, and estimate the parameters of the distribution.

> They compare a series of product-preference models to a pair of models that assume no correlation between "related" brands. That is, preference for brand j gives no information about the relative preference for brands m and n even though they may share similar characteristics. These simpler "Independence of Irrelevant Alternatives" models tended to perform as well as or better than any of the other models.

Epple, D., R. Romano, and H. Sieg. 2006. "Admission, Tuition, and Financial Aid Policies in the Market for Higher Education." *Econometrica* 74 (4) (July): 885–928.

This study does not look much like any other study in the Hedonic literature, but it solves essentially the same problem. They implicitly used the non-linearity of price to estimate preferences for higher education, as well as the cost-function for providing higher education. They explicitly develop a supply-demand model for higher education. Then they estimate the parameters in the model using a two-loop iterative procedure. The outer loop was a maximum likelihood estimation of parameters for preferences and costs. The inner loop used the candidate parameters for preferences and costs to solve for the equilibrium price for all the different levels of educational quality in the model. Their model gave rise "to a strict hierarchy of colleges that differ by the educational quality provided to the students." Based on their model they were able to "simultaneously [predict] student selection into institutions of higher education, financial aid, educational expenditures, and educational outcomes." However, the model was very computationally intensive. They had to use bootstrapping to estimate errors and each bootstrap iteration took 24 hours on a PC to compute.

Kristofersson, Dadi, and Kyrre Rickertsen. 2004. "Efficient Estimation of Hedonic Inverse Input Damand Systems." *American Journal of Agricultural Economics* 86 (4) (November): 1127–1137.

Kristofersson, D., and K. Rickertsen. 2007. "Hedonic Price Models for Dynamic Markets." *Oxford Bulletin of Economics and Statistics* 69 (3): 387–412.

These studies estimate the demand for fish on Iceland's fish auction markets. They are studying an intermediate market, and the customers whose preferences make up demand are fish processing firms. They used price information from multiple markets, except in their case the multiple markets are auctions that occur on different days. They assumed that supply in each daily market was inelastic, since the fish on each days market had already been caught. That allowed them to avoid a simultaneous equations model. They also developed a random-coefficients model for demand. In a random-coefficients approach, instead of estimating regression parameters, the researcher estimates a probability distribution over the parameters. They found that price per unit weight for Cod increased as the fish got larger. They were also able to estimate how much the price per unit weight declined per day of storage or if the fish was not gutted.

Palmquist, Raymond. 1994. "Estimating the Demand for the Characteristics of Housing." *The Review of Economics and Statistics* 66 (3) (August): 394–404.

This paper estimates demand and preferences for housing. He starts by estimating the price function for several different metropolitan areas. Instead of using the information to simultaneously estimate supply and demand, (like Witte et al. (1979), discussed below), he uses instrumental variables to identify the characteristics of demand.

Witte, A. D., H. J. Sumka, and H. Erekson. 1979. "An Estimate of a Structural Hedonic Price Model of the Housing Market: An Application of Rosen's Theory of Implicit Markets." *Econometrica: Journal of the Econometric Society*: 1151–1173.

This paper estimates preferences for rental housing and the cost function for provision of rental housing in non-metropolitan cities in North Carolina. They use price information from four geographically distinct markets in the state to estimate supply and demand parameters. They assume that there was a single parameterized utility function and cost function (with the parameters describing the difference in costs between different producers, as well as the difference in preferences between different consumers). First, they estimate the price function for each of the markets they analyzed. From that, they obtain marginal price information. Then they use the price information from multiple markets to estimate underlying preferences using standard multiple-equation techniques. Their supply and demand functions had the expected slope: that is, demand declined as its price increased and supply increased as its price increased. In demand, most parts of the composite good that formed the house were complements: that is, an increase in one good increased demand for the other that make up the house.

Price Function

Ahmad, Waseem, and Sven Anders. 2012. "The Value of Brand and Convenience Attributes in Highly Processed Food Products." *Canadian Journal of Agricultural Economics-Revue Canadienne D Agroeconomie* 60 (1) (March): 113–133.

> This is a standard hedonic regression. They look at prepackaged dinners for chicken and fish to estimate the monetary value of brand, convenience, and other quality attributes. The data they use is the 2006 Nielsen aggregate weekly scanner data. They "find evidence of consumer preferences for perceived natural and health attributes over products with higher degrees of processing."

Ambrose, Brent. 1990. "An Analysis of the Factors Affecting Light Industrial Property Valuation." *Journal of Real Estate Research* 5 (3) (January 1): 355–370.

> This is a straightforward estimation of the price functional for light industrial space. It is different from most studies in that it is applied to an intermediate good rather than a final good.

Boyle, Kevin J., Nicolai V. Kuminoff, Congwen Zhang, Michael Devanney, and Kathleen P. Bell. 2010. "Does a Property-specific Environmental Health Risk Create a 'Neighborhood' Housing Price Stigma? Arsenic in Private Well Water." *Water Resources Research* 46 (March 9).

> This study evaluates the impact on the price of housing of having arsenic in wells in the neighborhood. Their data set are housing prices from two towns in Maine where the existence of arsenic in well water was well-publicized. They found that housing prices suffered a temporary two-year decline following media coverage of the arsenic contamination, after which prices recovered.

Carew, Richard, Wojciech J. Florkowski, and Elwin G. Smith. 2012. "Hedonic Analysis of Apple Attributes in Metropolitan Markets of Western Canada." *Agribusiness* 28 (3): 293–309.

> This is a standard hedonic analysis of wholesale apple prices in western Canada. They estimate the price premiums paid for newer varieties, higher grades, and larger fruit size, as well as the price effects of cold-storage and seasonality.

Carter, David W., and Christopher Liese. 2010. "Hedonic Valuation of Sportfishing Harvest." *Marine Resource Economics* 25 (4): 391–407.

> This is an application of Hedonic methods to sportfishing charters in the Gulf of Mexico.

Deligiorgi, C., C. Michalakelis, A. Vavoulas, and D. Varoutas. 2007. "Nonparametric Estimation of a Hedonic Price Index for ADSL Connections in the European Market Using the Akaike Information Criterion." *Telecommunication Systems* 36 (4) (December): 173–179.

This study estimates the characteristics that determine price in Asynchronous DSL lines in Europe. They allowed the price to be a non-linear function of characteristics, and found that price was strongly related only to downlink data rate. They allow the relationship between price and download speed to be non-linear by using Sliced Inverse Regression.

Donovan, Geoffrey H., and David T. Butry. 2010. "Trees in the City: Valuing Street Trees in Portland, Oregon." *Landscape and Urban Planning* 94 (2) (February 28): 77–83.

This paper uses housing prices to estimate the value of having trees on the street. They take spatial correlation into account, and simultaneously estimate time on market. On average, street trees add $8870 to sales price and reduce time on the market by 1.7 days. In addition, they found that the benefits of street trees spill over to neighboring houses.

Fuerst, Franz, and Pat McAllister. 2011. "Eco-labeling in Commercial Office Markets: Do LEED and Energy Star Offices Obtain Multiple Premiums?" *Ecological Economics* 70 (6) (April 15): 1220–1230.

This is a fairly straightforward hedonic regression on the market for rented commercial space. They look both at the rental market (rent / unit of floor space) and at the sales market for the same set of buildings. They find that an energy efficient rating increases rental premia by about 3 – 5 %. Dual certification increases rents additively. Sales price increases 18 – 25 % for a single certification wile dual certification increases sales price 28 – 29%.

For their rental model, they use occupancy rate as one of the explanatory characteristics. However, occupancy is codetermined with price as the balancing of supply and demand. By including occupancy as one of the covariates they introduce a potential endogeneity bias that they fail to correct for.

Gallaugher, JM, and YM Wang. 2002. "Understanding Network Effects in Software Markets: Evidence from Web Server Pricing." *MIS Quarterly* 26 (4) (December): 303–327.

This paper estimates the characteristics that affect price in server software. They are particularly interested in determining whether network effects impacted price. A "network effect" would exist if server software with a large customer base is more valuable to customers *because* it has a large customer base. For example, a large customer base may inherently be associated with a large network of third-party support firms. They try to determine whether having a large market share (which presumably includes an associated network affect) contributed to the price of server software. They found that a network effect existed. Unfortunately, market share is endogenous: how many customers buy a particular suite of server software is itself a function of its price (as well as the characteristics of software). They did not account for the endogeneity, and that could bias their results.

Goodman, A. C. 1983. "Willingness to Pay for Car Efficiency: A Hedonic Price Approach." *Journal of Transport Economics and Policy*: 247–266.

This study estimates the value of fuel efficiency for cars on the used car market. He specifically looked at a select subset of models, sold on the used car market in 1977 and 1979, and that were two years old at the time. In 1977 he found that a one unit increase in fuel efficiency increased the resale price of 1975-model cars by approximately $93. However, in 1979, he found that an increase in fuel efficiency *reduced* the resale price of a 1977-model car. In spite of valiant effort, he was unable to overturn that conclusion. That result is counter-intuitive, since fuel efficiency would seem to be a good (as further indicated by the 1977 results).

These results are probably explained by supply-related factors. The Corporate Average Fuel Economy (CAFE) standards came into force in 1978 and were significantly tightened for the 1979 model year. Those standards require auto makers to maintain a minimum average fuel economy for the automobiles sold in a specific year. Assuming (as seems likely) that the standards mandated a mix of cars different from the one that consumers would have purchased on their own in 1979, then less fuel efficient new cars would have carried a supply-related surcharge. Since that increase in price did not affect demand (and assuming that new and used cars are to some extent substitutes), the "overpricing" of less fuel efficient new cars would to some extent have been compensated for by a similar "overpricing" of less fuel efficient used cars.

Halstead, J. M., R. A. Bouvier, and B. E. Hansen. 1997. "On the Issue of Functional Form Choice in Hedonic Price Functions: Further Evidence." *Environmental Management* 21 (5): 759–765.

This study estimates the impacts on housing value of being near a closed landfill. They found that proximity to the landfill had no impact on property values. What distinguishes this study from others is their effort to identify the functional form of the price function. They use the Box-Cox transformation, using different transformation coefficients on different variables. They estimate the optimal Box-Cox coefficient using maximum likelihood techniques.

Hough, Douglas E., and Charles G. Kratz. 1983. "Can 'good' Architecture Meet the Market Test?" *Journal of Urban Economics* 14 (1) (July): 40–54.

This paper estimates the price-characteristics relationship for office space in Chicago. In particular, they are interested in determining whether there was a price premium for "good" architecture. They defined "good" architecture as buildings that had been declared national landmarks or Chicago landmarks for architectural reasons, or buildings that had won an award for architectural excellence from the Chicago-American Institute of Architects. They found that "good" architecture in newer buildings carried a price premium, while "good" architecture in older buildings did not.

Hui, E., C. K. Chau, L. Pun, and M. Y. Law. 2007. "Measuring the Neighboring and Environmental Effects on Residential Property Value: Using Spatial Weighting Matrix." *Building and Environment* 42 (6): 2333–2343.

Hui, E., J. W. Zhong, and K. H. Yu. 2012. "The Impact of Landscape Views and Storey Levels on Property Prices." *Landscape and Urban Planning* 105: 86–93.

These are standard studies of housing prices for Hong Kong, in densely populated high-rise neighborhoods. They are particularly interested in the impact of having a sea view and better air quality on housing price. They also look at the impact of story, noise level, and proximity to green-belt spaces on price. They specifically take spatial correlation into account. For anyone interested in spatial correlation, these studies are a good place to start.

Jani, A. B., and S. Hellman. 2008. "Early Prostate Cancer: Hedonic Prices Model of Provider-patient Interactions and Decisions." *International Journal of Radiation Oncology Biology Physics* 70 (4) (March 15): 1158–1168.

This paper applies a version of Hedonic prices to determine what factors are most influential in the choices of therapy for Prostate Cancer.

Lavee, Doron, Tomer Ash, and Gilat Baniad. 2012. "Cost-benefit Analysis of Soil Remediation in Israeli Industrial Zones." *Natural Resources Forum* 36 (4) (November): 285–299.

This is a cost-benefit analysis of remediating all soil-contamination sites in Israel. They use a hedonic approach to estimate the benefits.

Liu, Jin-Long, Pe-I. Chang, and Su-Juan Den. 2012. "Consumer Willingness to Pay for Energy Conservation: a Comparison Between Revealed and Stated Preference Method." *Procedia Environmental Sciences* 17: 620-629.

This is a standard Hedonic model estimating the impact of an Energy Efficiency rating on the price of air conditioners in Taiwan.

Nguyen, Thong Tien. 2012. "Implicit Price of Mussel Characteristics in the Auction Market." *Aquaculture International* 20 (4) (August): 605–618.

This is a standard hedonic model of mussel values in a wholesale market in the Netherlands. He finds that meat content and size count, are the most important characteristics determining the price. He finds that impurity of mussel lots is a significant discounting factor on the price.

Rao, HR, and BD Lynch. 1993. "Hedonic Price Analysis of Workstation Attributes." *Communications of the ACM* 36 (12) (December): 95–102.

This is a standard hedonic model of workstation prices. By workstation, they mean the high-powered computer that was in between the Personal Computer and the Minicomputer in power. They found that the variables with the greatest effect on workstation price were the maximum size hard drive, whether or not the workstation came with a SCSI bus, and minimum RAM.

Robst, John. 2006. "Estimation of a Hedonic Pricing Model for Medigap Insurance." *Health Services Research* 41 (6) (December): 2097–2113.

This paper uses hedonic methods to estimate the characteristics that affect price of Medi-Gap insurance plans. His study differs from the others in that he is primarily interested in understanding the supply side of the market rather than the demand side. However, his methodology is essentially the same, and his data only allows the relationship of characteristics to price to be inferred.

Sirmans, Stacy, David Macpherson, and Emily Zietz. 2005. "The Composition of Hedonic Pricing Models." *Journal of Real Estate Literature* 13 (1) (January 1): 1–44.

This is a review of hedonic studies of housing. They are mainly interested in identifying consistently significant variables from those studies. It is a good round-up of the possible variables that could be used, and their expected sign and significance. It is a summary of their earlier study published by The National Association of Realtors.

Stetler, Kyle M., Tyron J. Venn, and David E. Calkin. 2010. "The Effects of Wildfire and Environmental Amenities on Property Values in Northwest Montana, USA." *Ecological Economics* 69 (11) (September 15): 2233–2243.

This study employed the hedonic price framework to examine the effects of wildfires and environmental amenities on home values in northwest Montana. They found that some environmental amenities, including proximity to lakes, golf courses, national forests, and Glacier National Park increased the value of properties. They also found that proximity to wildfire burn areas had a negative effect on price.

Von Auer, L., and M. Trede. 2012. "The Dynamics of Brand Equity: a Hedonic Regression Approach to the Laser Printer Market." *Journal of the Operational Research Society* 63 (10) (October): 1351–1362.

This is a hedonic analysis of the laser printer market in Germany. They are primarily interested in the price premiums associated with different brand names, and how those premiums vary over time. To account for the variation over time, they assume that the brand premium follows a random walk, and estimate the path it follows using Bayesian techniques.

Yu, Danlin. 2007. "Modeling Owner-occupied Single Family House Values in the City of Milwaukee: A Geographically Weighted Regression Approach." *GIScience & Remote Sensing* 44 (3): 267–282.

This study estimated housing prices for the city of Milwaukee, Wisconsin. It used a subset of the standard parameters used in housing. What distinguished this study from others like it is that it allowed prices to differ depending on location. It used a technique called "Geographically Weighted Regression" to estimate a non-parametric function of price as a function of location for each hedonic characteristic. It found that price for characteristics varied from location to location. For example, price for floor space varied from a low of $18 per square foot ($194 / m^2) to a high of $120 per square foot ($1,290 / m^2) in different parts of the city.

Zietz, Joachim, Emily Zietz, and G. Sirmans. 2008. "Determinants of House Prices: A Quantile Regression Approach." *The Journal of Real Estate Finance and Economics* 37 (4): 317–333.

This is a hedonic study of housing values in Orem/Provo Utah. They explore the possibility that the value of some characteristics may vary non-linearly. To examine this issue, they use quantile regression. They estimate the coefficients on housing characteristics for several different quantiles. "The results show that purchasers of higher-priced homes value certain housing characteristics such as floor area and the number of bathrooms differently from buyers of lower-priced homes. Other variables such as age are also shown to vary across the distribution of house prices."

www.ingramcontent.com/pod-product-compliance
Lightning Source LLC
Chambersburg PA
CBHW081414170526
45166CB00010B/3340